Bibliographic information published by the German National Library:

The German National Library lists this publication in the National Bibliography; detailed bibliographic data are available on the Internet at http://dnb.dnb.de .

Imprint:

Print and binding: Books on Demand GmbH, Norderstedt Germany
ISBN: 9783668370708

This book at GRIN:

http://www.grin.com/en/e-book/349822/swarming-drone-movements-foraging-behavior-and-associated-plants-of-stingless

Prem Jose Vazhacharickal, Sajan Jose K.

Swarming, drone movements, foraging behavior and associated plants of stingless bees (Trigona iridipennis Smith) in Kerala

An overview

GRIN Publishing

Swarming, drone movements, foraging behavior and associated plants of stingless bees (*Trigona iridipennis* Smith) in Kerala: an overview

Prem Jose Vazhacharickal and Sajan Jose K

ACKNOWLEDGEMENT

Firstly we thank **God Almighty** whose blessing were always with us and helped us to complete this research work successfully.

The first author is extremely grateful to **Dr. Sajeshkumar N.K** (Head of the Department, Biotechnology) for the valuable suggestions, support and encouragements.

The first author would wish to thank beloved Manager **Rev. Fr. Dr. George Njarakunnel,** respected Principal **Dr. V.J. Joseph,** Bursar **Shaji Augustine,** Vice Principal **Fr. Joseph Allencheril**, and the Management for providing all the necessary facilities in carrying out the study.

We lovingly and gratefully indebted to our teachers, parents, siblings and friends who were there always for helping us in this project.

Prem Jose Vazhacharickal and Sajan Jose K

Table of contents

Table of figures

Table of tables

List of abbreviations

AC	: Field crops
CS	: Condiments and spices
FC	: Fruit crops
M	: Medium
MP	: Medicinal plants
N	: Nectar
NP	: Nectar and pollen
OP	: Ornamental plants
P	: Pollen
PC	: Plantation crops
S	: Strong
SPSS	: Statistical package for social sciences
ST	: Shade trees
TR	: Trees
VC	: Vegetable crops
W	: Weak

Swarming, drone movements, foraging behavior and associated plants of stingless bees (*Trigona iridipennis* Smith) in Kerala: an overview

Prem Jose Vazhacharickal[1]* and Sajan Jose K[2]

* premjosev@gmail.com

[1]Department of Biotechnology, Mar Augusthinose College, Ramapuram, Kerala, India-686576

[2]Department of Zoology, St. Joseph's College, Moolamattom, Kerala, India-685591

Abstract

Stingless bees are highly social insects which populated the tropical earth 65 million years ago longer than honey bees. They are limited to tropics and subtropics lacking venom apparatus and cannot sting. Impacts of anthropogenic influences on honey bees were already reported. Based on these back ground, the objectives of this study were to 1) to characterize the swarming and drone movements of *Trigona iridipennis* Smith in Kerala 2) to identify the foraging behaviour, colour preference and various plants associated with *Trigona iridipennis* Smith for nest construction as well as dietary sources. Feral as well as hived colonies of stingless bees were located and fixed at different parts of Kerala for observations on their foraging activity. Bee traffic was also measured during various timings across a bright sunny day in November 2013 and between colonies. Bee traffic among *Trigona iridipennis* Smith varied widely among the selected four colonies during various time intervals. In all the colonies bee traffic starts between 6 to 7 am and end by 7 to 7.10 pm. Bee traffic exhibited two peaks. One during 10 to 10.10 am and the other 2 to 3.10 pm. But in colony 4 the picture is different, where the morning peak was 11 to 11.10 am and the afternoon peak was 3 to 3.10 pm. *Trigona iridipennis* Smith shows great diversity in plant selection for dietary as well as resin sources. The shift towards ornamental plants for foraging may be an adaptation evolved in response to human modification of the environment. The bees collect resin from a variety of sources for building nest, its maintenance and also for defence. Bee traffic is found to be related to time, season, and strength of the colony. The bees preferred white and yellow coloured flowers than pink and red. The study also highlights the various food sources of *Trigona iridipennis* Smith in Kerala which can be further explored for flourishing melliponiculture.

Keywords: *Trigona iridipennis* Smith; Anthropogenic habitats; Entrance tube; Meliponiculture; Colour preference; Swarming.

1. Introduction

Stingless bees populated tropical earth for over 65 million years before present and much longer than Apis, the stinging honey bees (Camargo and Pedro, 1992; Michener, 2000). Both groups make honey in perennial nests founded by a swarm or sterile workers and a queen, and colonies occasionally produce male bees. Yet stingless bees have 50 times more species and differ from Apis in many biologically significant ways. Meliponines cannot migrate and unlike honey bees, they produce brood in the manner of solitary bees, with an egg placed on top of a food mass in a sealed cell. In general, colonies make far less honey, and therefore have less economic appeal, compared to honey bees (Roubik, 2006). Meliponines have vestigial sting, mate only once in life time, or like the Apis they do not use water to cool their nest or pure wax to build it, cannot freely swarm to reproduce (but instead must first make a new domicile). The males feed at flowers and the gravid queens cannot fly. It is from the nest they move out for mating, foraging and various life activities. Nests are usually immobile and potentially long lasting (Roubik, 2006).

Stingless bees are eusocial bees living in walls, trees, crevices, and such other concealed places. They belong to the family Apidae and sub family Meliponinae. Meliponinae consists of two genera Melipona and Trigona which belong to the tribe Meliponini and Trigonini, respectively. Meliponinae includes eight genera, having fifteen subgenera and more than 500 species (Wille, 1983). Stingless bees are most diverse, morphologically and behaviourally, among the eusocial corbiculate bees (Michener, 2000). Facets of their diversity are evident in their social organization, systems of communication, nest architecture and reproductive behaviour. Their perennial colonies range in size from fewer than 100 to tens of thousands of workers (Roubik, 1989; Michener, 2000) and usually contain a single queen (Velthuis et al., 2001). Meliponines are distributed throughout the tropical and subtropical parts of the Afrotropical, Australasian, Indo-Malayan, and Neotropical regions, exhibiting greatest abundance in New World Amazonian rain forest (Michener, 1979; Roubik, 1992; Michener, 2000).

With few exceptions, both female bees and males gather food from flowers. In addition, except for those that usurp the nest of others, most bees gather nesting material and provisions for their larvae. Their resource menu is comparatively larger for the existence and survival of the colony. Foraging bees normally seek plant products such as gums, resins, rotten wood, bark, fruit juices, seeds, leaves, plant hairs or trichomes, fragrances, pollen, nectar, oil, spores and rusts, sap, and the honeydew excreted by plant-feeding homopteran bugs and fungi, other natural products such as wax, animal faeces, carrion, urine, and hairs, and combined plant and animal products such as cerumen and propolis. Other materials used by bees include mud, loose soil, gravel, various salt solutions, and water (Roubik, 1989).

Stingless bees (Meliponini) show a pantropical distribution and are important pollinators of melittophilous plants in tropical rain forests (Kerr and Maule 1964, Michener 1979, Roubik, 1989). Stingless bees are widely known as “dammer bees” or “dammar bees” as they collect resin from amongst dipterocarp trees. In Kerala stingless bees are known as “cherutheneecha” and “arakki” (Nair, 2003). Stingless bees are known from most parts of the Indian subcontinent, at least up to 1000 m AMSL (above mean sea level) in India and Nepal and they are also called mosquito bees (Sinu and Shivanna 2007, Sing, 2013).

The name *Trigona* refers to their triangular abdomen and ‘*iridipennis*’ refers to their iridescent wings. Stingless bees are small to medium sized with vestigial stings, shows social level of organization and pollinators of flowering plants in the tropics (Ramanujam et al., 1993; Heard, 1999; Amano et al., 2000; Raju et al., 2009). The stingless bee nest is always characterized by a nest cavity, typically provided with a very narrow opening facilitating defence (Kolmes and Sommeijer, 1992).

Trigona spp. has widespread distribution over the tropical and subtropical areas of the world. They are valuable pollinators of many crops (Heard, 1999; Devanesan et al., 2004; Viraktamath et al., 2013; Jayarathnam, 1970). The female possess vestigial stingers and are unable to inflict pain with them, hence the term “stingless bees”. Some species have strong mandibles sufficient to inflict mild bite and they may crawl into the ears and nostrils of the intruders. Others emit a caustic liquid from the mouth that, in contact with the skin, causes intense irritation (Greco et al., 2010; Lehmberg et al., 2008; Smith and Roubik, 1983). Smaller bees have lower metabolic costs, and can profitably forage on flowers providing lower rewards compared to larger bees (Corbet et al., 1995).

A little is reported so far about the various behaviours of the *Trigona iridipennis* Smith as well as their foraging activity in Kerala. According to Singh (2013) and Mohan and Devanesan (1999), the nest of *Trigona iridipennis* Smith consist of entrance, brood, food storage pots, resin dump, waste dump, pillars, connectives, involucrum, and batumen. Difference in the entrance depending on the species (Chinh et al., 2005; Bänziger et al., 2011; Pavithra et al., 2012; Danaraddi et al., 2012) as well as tree preference for building nest (Marisa and Salni, 2012). Based on these back ground, the objectives of this study were to 1) to characterize the swarming and drone movements of *Trigona iridipennis* Smith in Kerala 2) to identify the foraging behaviour, colour preference and various plants associated with *Trigona iridipennis* Smith for nest construction as well as dietary sources.

2. Materials and Methods

2.1 Study area

Kerala state covers an area of 38,863 km^2 with a population density of 859 per km^2 and spread across 14 districts. The climate is characterized by tropical wet and dry with average annual rainfall amounts to 2,817 ± 406 mm and mean annual temperature is 26.8°C (averages from 1871-2005; Krishnakumar et al., 2009; Figure 1). Maximum rainfall occurs from June to September mainly due to South West Monsoon and temperatures are highest in May and November.

2.2 Study design and data collection

Feral as well as hived colonies of stingless bees were located and fixed at different parts of Kerala for observations on their foraging activity. Acceptance of a flower was recorded when the bee landed on the flower and gathered pollen or probed for nectar. Rejection was recorded if the bee approached to within 1 cm of the flower and then departed without gathering pollen probing for nectar (Goulson et al., 2001). Field observations were made to find out the pasturage sources that provide pollen, nectar and resin. The plant was identified as a nectar source if the bee calmly sits on the flower to sip the nectar. The plant was judged as a pollen source if the forager bee restlessly moves on the flower and the pollen basket is filled with pollen when it returns. When the bee exhibits both the behaviours the plant was considered as a source of both pollen and nectar. The plant was considered as a resin source when the corbiculae got filled with sticky resin when it returns.

Bee traffic was also measured during various timings across a bright sunny day in November 2013 and between colonies. Four colonies of average strength kept in wooden boxes were selected for this study. The number of workers leaving the nest and the number of workers entering into the nest with pollen or nectar load was recorded for a period of ten minutes at one hour intervals from 06:00 to 19:10 hrs.

Drone movements and swarming experiments were conducted with an exception in the number of observed colonies. Twenty stingless bee colonies were continuously monitored for a period of twelve months. Drone movements were identified as the unusual movement of drone bees with head directed towards nest entrance and a few of them may rest on the nearby surfaces. These bees were collected using bee trap net (2 mm) and were later analyzed to confirm their gender. Swarming were identified as the group movement of bees which may either aggressive or non aggressive and lasts up to 3 to 4 days. Aggressive swarming was identified as irregular movements of worker bees in front of the nest, clutching together of two bees and their subsequent falling to the ground and death. A normal swarming is completely different from the above mentioned phenomena where a fast circular movement of bees observed in front of the nest for a few days.

Net log experiments were conducted to determine the fissure closure ability of the stingless bees in domesticated hives. Six colonies in wooden boxes were selected for this experiment. Top lid of the colony was replaced using polyethylene net with a mesh size of 2 mm. The time for complete closure of holes in net was considered as the fissure closure ability.

Colour preference studies among stingless bees were conducted in four different gardens (garden 1, 2, 3 and 4) where single stingless bee colony kept. Two plant variety; rose-moss (*Portulaca grandiflora;* garden 1 and 2) and Ixora (*Ixora coccinea;* garden 3 and 4) with four different flower colours (white, yellow, pink and red) were observed in 8 consecutive days (day 1 to 8). Flower colour preference was recorded based on the maximum visitation of the bees on a particular flower.

Drone movements and swarming experiments were carried out by monitoring twenty colonies continuously for a period of twelve months. Drone movements were identified as the usual movement of drone bees with head directed towards nest entrance and few of them found to rest on nearby surfaces. These bees were collected using trap nest (2 mm) and were later analyzed to confirm their gender. A normal swarming is identified as group movements of bees which may lasts up to 3 to 4 days. It is a type of fast circular movement of bees in front of the nest (Michener, 1974; Terada, 1975; Inoue et al., 1984; Van Veen et al., 1997; Peters et al., 1999). The bees also exhibit aggressive swarming, which is identified as the irregular movements of worker bees in front of the nest, clutching together of two bees and their subsequent falling to the ground and death (Hubbel and Johnson, 1978; Breed and Page, 1991).

2.3 Statistical analysis

Descriptive statistics using SPSS 12.0 (SPSS Inc., Chicago, IL, USA) were conducted to summarize the data and graphs were generated using Sigma Plot 7 (Systat Software Inc., Chicago, IL, USA).

3. Results and discussion

3.1 Pasturage sources

Stingless bees collect pollen to provision their nest with food for the rearing of brood. Pollen thieving by these bees is maximised, however, in the absence of shade, a condition which also denotes the activity of pollinators owing principally to thermal and moisture stress (Young, 1983). More than 90% of the pollen collection flights are made between 5:00 and 9:00 hrs. Colonies show a gradual reduction in pollen collection during day. Nectar foraging flights shows a more uniform distribution during the day (Pierrot 2003). In favourable season foraging took place throughout the day but it was most active between 11:00 and 13:00 hrs (Giovannini 1986). A peak flight activity observed at 28 °C (Giovannini 1986).

The major identified plants were grouped into plantation crops, medicinal plants, ornamental plants, vegetables, fruit crops, condiments and spices, fruit crops, trees and shade trees (Table 1 to 9). The preferences showed by the bees for the different plants were in the order; ornamental plant (61)> medicinal plants (21)> fruit crops (16) ~ trees (15)> vegetables (14)> field crops (9)> plantation crops (8)> shade trees (6) ~ condiments and spices (5). It was found that the bees visited a total of 155 plants. The availability of diverse flora and the adaptability of the bees help them to forage many varieties. Majority of the plants belong to the family Arecaceae, Euphorbaceae, Asteraceae and Cucurbitaceae.

The workers from a stingless bee colony can visit different types of plants. This behaviour, called polylecty (Goulson et al., 2001), enables a colony to potentially pollinate variety of plants. In addition, they can also quickly adapt to new plants that they have not known before. Each individual worker on a trip usually visits only one plant species. This behaviour, called floral constancy, makes these bees efficient pollinators because each bee only carries pollen between the flowers of one plant species (Roubik, 1995). *Trigona iridipennis* shows great diversity in plant selection for dietary as well as resin sources. Here the most preferred category of plants is the ornamentals. The shift towards selecting ornamental flowers as their food source may be considered as a secondary adaptation to human habitats (Devanesan et al., 2009).

Table 1. Plantation crops as source of nectar and pollen for *Trigona iridipennis* Smith in Kerala

Serial number	Common name	Scientific name	Family	Source*
1	Coconut	*Cocos nucifera* L.	Palmae	NP
2	Arecanut	*Areca catechu* Linn.	Palmae	P
3	Tea	*Camellia sinensis*	Camelliaceae	P
4	Cashew	*Anacardium occidentale*	Anacardaceae	NP
5	Coffee	*Coffea arabica* L.	Rubiaceae	N
6	Rubber	*Hevea brasiliensis*	Euphorbiaceae	N
7	Oil palm	*Elacis guineensis*	Aracaeceae	P
8	Eucalyptus	*Eucalyptus sp.*	Myrtaceae	NP

* N= nectar, P= pollen, NP= nectar and pollen

Table 2. Medicinal plants as a source of nectar and pollen for*Trigona iridipennis* Smith in Kerala

Serial number	Common name	Scientific name	Family	Source*
1	Touch-me-not	*Mimosa pudica*	Mimosaceae	P
2	Thulsi	*Osmium sanctum*	Laminaceae	N
3	Puliyarila	*Oxalis carniculata*	Oxalidaceae	N
4	Ixora	*Ixora coccinea*	Rubiaceae	NP
5	Henna	*Lowsonia alba*	Lythraceae	P
6	Castor .	*Ricinus communis*	Euphorbiaceae	N
7	Neem	*Azadirachta indica*	Meliaceae	N
8	Nagadandi	*Baliospermum monatanum*	Euphorbiaceae	N
9	Thazhuthama	*Boerhavia diffusa*	Nyctaginaceae	N
10	Parijathum	*Nyctanthes arbortristis*	Oleaceae	N
11	Phyllanthus	*Phyllanthus niruri*	Euphorbaceae	N
12	Periwinkle	*Vinca rosea*	Apocynaceae	N
13	Tridax	*Tridax procumbens*	Compositae	NP
14	'Kallurukky'	*Scoparias dulce*	Scrophulariaceae	N
15	Sesbania	*Sesbania rostrata*	Papilionaceae	N
16	'Neela amari'	*Indigofera tinctoria*	Fabaceae	P
17	Trumpet plant	Brugmansia suaveolens	Solanaceae	P
18	Javanese wool plant	*Aerva lanata*	Amaranthaceae	N
19	Kurumthotti	*Sida cordiflorus*	Malvaceae	NP
20	Gladiolus	*Gladiolus grandiflorus*	Iridaceae	NP
21	Thumba	*Leucas aspera*	Lamiaceae	NP

* N= nectar, P= pollen, NP= nectar and pollen

3.2 Pollen source

In bees pollen foraging in the morning is more intense than in the afternoon. Nectar collection is more intense in the late morning hours and afternoon. Resin collection is evenly distributed between morning and afternoon and mud was collected more frequently in the morning. Waste is removed more frequent in the morning (Pierrot and Schlindwein, 2003). The various pollen sources for stingless bees in Kerala were given in table 1 to 9. The major identified plant families were Palmae, Fabaceae, Mimosaceae and Aracaeceae.

3.3 Nectar source

Stingless bees can collect nectar from small flowers due to their small size. Various nectar sources for *Trigona iridipennis* Smith were presented in table 1 and 9. Colour preferences and attraction were observed in stingless bees for selecting flowers. Among the different flowers, colour preferences were white followed by yellow, pink and red respectively. The major identified plant families were Amaranthaceae, Papilionaceae, Apocynaceae and Euphorbaceae.

3.4 Resin source

Stingless bees collects resins from a variety of plant species including trees, herbs and shrubs. Various resin sources for *Trigona iridipennis* Smith were presented in Table 10. The resin is collected from resinous plants like Jack fruit, mango tree etc. and wax is secreted by the young worker bees. The chemistry of propolis depends on the diversity of plants from which the bees collect it (Pereira, 2003).

3.5 Resin collection procedure

Resin is used for building nest, brood, honey pot, pollen pot, pillars, connectives, entrance tube and internal tunnel. Resin mixed with wax (cerumen) is used in building various structures mentioned above but their composition may vary. Resin is collected from various woody and non-woody plant sources which include latex, gum and various resins. Resin collection is a risky task which may even leads to the entrapment of worker bees inside the sticky resin. Only experienced elder bees take care of this complicated task due to its risk.

Worker bees locate the resin sources especially wounded tree barks including cut tree branches, and slowly bite the sticky resin using its strong mandibles. The collected resins were rolled into small balls using forelimbs which later transferred to the corbicula/pollen basket in the hind limbs. Many of the elder worker bees may get trapped in resin while doing this risky operation.

3.6 Net-log experiment

Net-log experiments revealed the fissure closure ability of the stingless bees in domesticated hives. The minimum time for the closure was 36 hrs while the maximum was 72 hrs. Strong colonies showed high fissure closing ability while weak colonies required more time for the complete closure.

3.7 Nesting trees

Trigona iridipennis Smith is seen in forest/wild as well as in natural habitats. The major natural habitats include the living trunks of different tree species especially teak (*Tectona grandis* Linn), jack fruit (*Artocarpus heterophyllus* Lam.), wild jack fruit (*Artocarpus hirsutus* Lam.), cycas (*Cycas sphaerica*) and mango (*Mangifera indica* L). Usually *Trigona iridipennis* Smith occupies in tree trunks of diameter above 30 cm and levels of few centimetres above the ground. They build the nests on trees which have hollow spaces due to decay or rotting of the stem or braches which arise due to insect attack, disease as well as physical damage. In central Travancore majority of the nests in the tree trunks was seen on "Maruthu" (*Terminalia paniculata*) and at coastal areas it was "Poovarasu" (*Hopea glabra*). It may be because both these trees have lot of cavities in their trunk, as farmers cut their branches regularly for mulching.

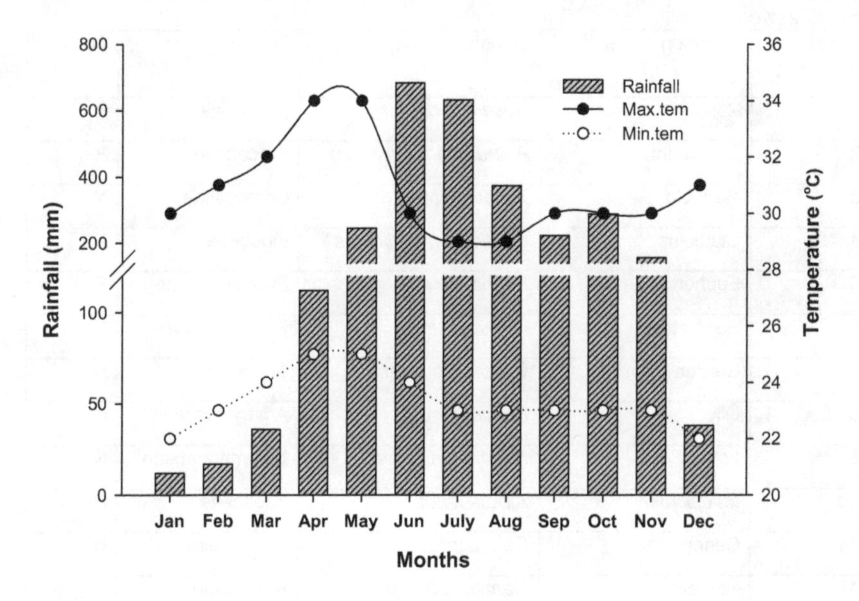

Figure 1. Mean monthly rainfall (mm), maximum and minimum temperatures (°C) in Kerala, India (1871-2005; Krishnakumar et al., 2009).

Table 3. Ornamental plants as a source of nectar and pollen for*Trigona iridipennis* Smith in Kerala.

Serial number	Common name	Scientific name	Family	Source*
1	Rose	*Rosa sinensis* L.	Rosaceae	P
2	Anthurium	*Anthurium andreanum*	Arecaceae	P
3	Marigold	*Tagetes erecta*	Compositae	N
4	Gladiolus	*Gladiolus grandiflorus*	Iridaceae	N
5	Euphorbia	*Euphorbia milii*	Euphorbiaceae	P
6	Honey tree	*Pittosporum sp*	Pittosporaceae	N
7	Garden palm	*Cyrtostachys renda*	Arecaceae	N
8	Lilly	*Pancratium sp*	Amaryllidaceae	P
9	Lotus	*Nelumbo nucifera*	Nelumbonaceae	N
10	Manja vaka	*Albizia lebbeck*	Fabaceae	P
11	Canna	*Canna indica*	Cannaceae	N
12	Hamelia	*Hamelia patens*	Rubiaceae	N
13	Balsum	*Impatiens balsaminae*	Balsaminaceae	N
14	Ball lilly	*Haemanthus cinnabarinus*	Amaryllidaceae	P
15	Bird of paradise	*Helicona rostrata*	Zingiberaceae	N
16	Bottle brush	*Callistemon speciosus* DC.	Myrtaceae	NP
17	Coral vine	*Antigonum leptopus*	Polygonaceae	NP
18	Dragon plant	*Dracaena fragrans*	Agavaceae	NP
19	Orchid	*Spathoglottis plicata*	Orchidaceae	P
20	Gerbera	*Gerbera sp.*	Asteraceae	N
21	Bauhinia	*Bauhinta racemosa*	Caesalpiniaceae	NP
22	Ross-moss	*Portulaca grandiflora*	portulacaceae	P
23	Cosmos	*Cosmos bipinnatus*	Asteraceae	N
24	'Venthi'	*Tagetus erectus*	Asteraceae	N
25	Carnation	*Dianthus caryophyllaceus*	Caryophyllaceae	N

26	Ixora	*Ixora coccinea*	Rubiaceae	N
27	Sunflower	*Helianthus annuum*	Asteraceae	N
28	Aerva	*Aerva lanata*	Amarantheceae	NP
29	Nymphea	*Nymphea stellata*	Nympheaceae	P
30	Peacock plant	*Caesalpinia pulcherima*	Caesalpiniaceae	NP
31	Golden dewdrop	*Canna indica*	Cannaceae	N
32	Poinsettia	*Euphorbia pulcherima*	Euphorbiaceae	N
33	Cosmos	*Cosmos sulfureus*	Asteraceae	NP
34	Yesterday-today	*Brunfelria calycinae*	Solanaceae	N
35	Lonicera	*Lonicera elaeagnoidea*	Caprifoliaceae	P
36	Celosia	*Celosia cristata*	Ameranthaceae	N
37	Murraya	*Murraya exotica*	Rutaceae	P
38	Blue	*Petria volubilis*	Verbenaceae	N
39	Euphorbia	*Euphorbia heterophylla*	Euphorbiaceae	N
40	Trumpet plant	*Brugmansia suaveolens*	Solanaceae	N
41	Portulaca	*Portulaca oleraceae*	Portulacaceae	NP
42	Duranta	*Duranta goldiana*	Verbinaceae	N
43	Gold spot	*Duranta plumieri*	Verbinaceae	N
44	Golden rod	*Solidago canadensis*	Compositae	P
45	Sage	*Salvia splendens*	Labiatae	P
46	Jamanthi	*Chrysanthemum sp.*	Asteraceae	NP
47	Coleostephus	*Coleostephus sp.*	Asteraceae	P
48	Celosia	*Celosia spicata*	Amaranthaceae	P
49	Hibiscus	*Hibiscus rosa-sinensis*	Malvaceae	P
50	Tristellateia	*Tristellateria australis*	Malpighiaceae	NP
51	Turnera	*Turnera subulata*	Turneraceae	NP
52	Gardenia	*Gardenta jasminoides*	Rubiaceae	NP
53	Murraya	*Murraya paniculata*	Rutaceae	NP
54	Ehretia	*Ehretia buxifolia*	Myrsinaceae	P

55	Pureria	*Pureria sp.*	Papilliionaceae	NP
56	Aster	*Celeosteplnus myconis*	Asteraceae	NP
57	Mimosa	*Mimosa sp.*	Mimosaceae	NP
58	Rivina	*Rivina humtlis*	Phytolaccaceae	NP
59	Urena	*Urena lobata*	Malvaceae	NP
60	Ixora	*Ixora parviflora*	Rubiaceae	N
61	Lily	*Ltlium candidum*	Lillaceae	NP

* N= nectar, P= pollen, NP= nectar and pollen

Figure 2. Stingless bee workers foraging on a) *Capsicum annum*, b) & d) *Albizia lebbeck*, c) *Averrhoa bilimbi*, e) *Chrysanthennum sp.*, f) *Coleostephus sp.*, g) *Abelmoschus esculentus*, h) *Antigonum leptopus*.

Table 4. Vegetable crops as a source of nectar and pollen for*Trigona iridipennis* Smith in Kerala.

Serial number	Common name	Scientific name	Family	Source*
1	Brinjal	*Solanum melongena L.*	Solanaceae	P
2	Bitter gourd	*Momordica charantia L.*	Cucurbitaceae	P
3	Drumstick	*Morinja oleifera*	Moringaceae	NP
4	Ash gourd	*Benincasa hispida*	Cucurbitaceae	P
5	Snake gourd	*Trichosanthes cucurmerina*	Cucurbitaceae	P
6	Curry leaf	*Murraya koenigii*	Rutaceae	P
7	Chilly	*Capsicum annum*	Solanaceae	NP
8	Sponge gourd	*Luffa cylindrical*	Cucurbitaceae	P
9	Lady's finger	*Abelmoschus esculentus*	Malvaceae	P
10	Bottle gourd	*Lagenaria vulgaris*	Cucurbitaceae	P
11	Capsicum	*Capsicum frutescens*	Solanaceae	NP
12	Sweet gourd	*Momordica cochinchinensis*	Cucurbitaceae	P
13	Cheera	*Amaranthus sp*	Amaranthaceae	P
14	Pumpkin	*Cucurbito pepo*	Cucurbitaceae	P

* N= nectar, P= pollen, NP= nectar and pollen

Table 5. Fruit crops as a source of nectar and pollen for*Trigona iridipennis* Smith in Kerala.

Serial number	Common name	Scientific name	Family	Source*
1	Mango	*Mangifera indica L*	Caesalpiniaceae	NP
2	Papaya	*Carica papaya L.*	Caricaceae	P
3	Guava	*Psidium guajava L.*	Myrtaceae	P
4	Pomegranite	*Punica granatum L.*	Punicaceae	P
5	Banana	*Musa paradisica L*	Musaceae	P
6	Jack fruit tree	*Artocarpus integrifolius*	Moraceae	P
7	Rambutan	*Nephelium lappapeum*	Sapindaceae	N
8	Rose apple	*Euginea jambosa*	Myrtaceae	NP
9	Loovi	*Flacourtia inermis*	Flacourtiaceae	N
10	Carambola	*Averrhoa carambola*	Oxalidaceae	P
11	Jamun	*Eugenia cumini*	Myrtaceae	N
12	Passion fruit	*Passiflora edulis Sims.*	Passifloraceae	NP
13	Gooseberry	*Emblica officinalis*	Euphorbiaceae	P
14	Cherry	*Carissa carandas*	Apocynaceae	N
15	Bilimbi	*Averrhoa bilimbi*	Oxalidaceae	N
16	Litchi	*Litchi chinensis*	Sapindaceae	NP

* N= nectar, P= pollen, NP= nectar and pollen

Table 6. Condiments and spices as a source of nectar and pollen for*Trigona iridipennis* Smith in Kerala.

Serial number	Common name	Scientific name	Family	Source*
1	Tamarind	*Tamarindus indica L.*	Caesalpiniaceae	N
2	Cardamom	*Elettaria cardamom*	Zingiberaceae	NP
3	Cinnamon	*Cinnamomum zeylanicum*	Lauraceae	P
4	Mustard	*Brassica juncea*	Umbellifera	N
5	Coriander	*Coriandrum sativum L.*	Umbelliferae	NP

* N= nectar, P= pollen, NP= nectar and pollen

Table 7. Field crops as a source of nectar and pollen for *Trigona iridipennis* Smith in Kerala.

Serial number	Common name	Scientific name	Family	Source*
1	Tapioca	*Manihot esculenta*		NP
2	Gingelly	*Sesamum indicum L.*	Pedaliaceae	N
3	Onion	*Allium cepa*	Liliaceae	NP
4	Cotton	*Gossypium hirsutm*	Malvaceae	P
5	Jute	*Corchous olitorlus*	Tiliaceae	P
6	Pigeon pea	*Cajanus cajan*	Papilionaceae	N
7	Sunflower	*Helianthus annus L.*	Asteraceae	NP
8	Castor	*Ricinus communis L.*	Euphorbiaceae	P
9	Jetropha	*Jetropha sp.*	Euphorbiaceae	N

* N= nectar, P= pollen, NP= nectar and pollen

Table 8. Trees as a source of nectar and pollen for*Trigona iridipennis* Smith in Kerala.

Serial number	Common name	Scientific name	Family	Source*
1	Fig	*Ficus roxburghii*	Moraceae	P
2	Cannon Ball Tree	*Cauropeta guineensis*	Lecythidaceae	NP
3	Cotton tree	*Bombax malabarium*	Malvaceae	P
4	Birds cherry	*Mundingia calbura*	Verbinaceae	NP
5	Sandal	*Santalum album*	Sandalaceae	NP
6	Payyani	*Paganelia longifolia*	Bignoniaceae	P
7	Soapnut	*Sapindus emarginatus*		NP
8	Teak	*Tectona grandis Linn.*	Verbanaceae	P
9	Coper pod tree	*Peltophorum roxburghii*	Caesalpinaceae	P
10	'Ettilamaram'	*Schefflera stellata*	Araliaceae	NP
11	Bilimbi	*Averrhoa bilimbi*	Oxalidaceae	N
12	Jack tree	*Artocarpus integrifolius*	Moraceae	P
13	Oil palm	*Elacis guineensis*	Aracaceae	P
14	'Nagappoomaram'	*Couropita guianensis*	Lecithidaceae	NP
15	Bombax	*Bombax malabarium*	Malvaceae	P

* N= nectar, P= pollen, NP= nectar and pollen

Table 9. Shade trees as a source of nectar and pollen for*Trigona iridipennis* Smith in Kerala.

Serial number	Common name	Scientific name	Family	Source*
1	Wild Tapioca	Manihot glaziovi	Euphorbiaceae	NP
2	Glyricidia	Glyricidia maculate	papilionaceae	N
3	Birds cherry	*Mundingia calbura*	Verbinaceae	NP
4	Agave	*Agave americana*	Agavaceae	NP
5	Coper pod tree	*Peltophorum roxburghii*	Caesalpinaceae	P
6	Elengi	*Mimus elengi*	Sapotaceae	NP

* N= nectar, P= pollen, NP= nectar and pollen

Table 10. Resin sources for*Trigona iridipennis* Smith in Kerala

Sl. No.	Common Name	Scientific Name	Family	Source of Resin
1	Mango tree	*Mangifera indica*	Caesalpiniaceae	Cut stem
2	Jack fruit tree	*Artocarpus heterophyllus* Lam.	Moraceae	Cut branches, fruit stalk
3	Tapioca	*Manihot utilissima*	Euphorbiaceae	Petiole
4	Drum stick	*Moringa oleifera*	Moringaceae	Stem
5	Banyan tree	*Ficus bengalensis*	Moraceae	Cut stem
6	Vatta	*Macaranga peltata*	Euphorbiaceae	Cut stem
7	Kudampuli	*Garcinia cambogia*	Clusiaceae	Cut stem
8	Cashew	*Anacardium occidentale*	Anacardaceae	Cut stem
9	Sun flower	*Helianthus annus*	Asteraceae	Flower bud
10	Christmas tree	*Arecaria exelsa*	Arecaceae	Cut stem
11	Fig tree	*Ficus roxburgii*	Moraceae	Cut stem
12	Wild jack fruit tree	*Artocarpus hirsutus* Lam.	Moraceae	Cut branches, fruit stalk
13	Breadfruit tree	*Artocarpus altilis*	Moraceae	Cut branches, fruit stalk

Figure 3. Stingless bee worker foraging on a) *Portulaca oleraceae*, b) *Coleostephus myconis*, c) *Dianthes sp.*, d) *Pittosporum sp.*, e) & f) *Euphorbia milii*, g) *Ixora coccinea*, h) *Pancratium sp.*

Figure 4. Stingless bee worker foraging on a) *Celosia spicata*, b) *Helianthus annus*, c) *Mimosa sp.*, d) *Cocos nucifera*, e) *Hibiscus rosa-sinensis*, f) *Tristellateia australis*, g) *Turnera subulata*, h) *Anthurium andreanum*.

Figure 5. Stingless bee worker foraging on a) *Rivina humilis*, b) *Cauroupita guianensis*, c) *Euphorbia sp.*, d) *Aerva lanata*, e) *Theobroma cocoa*, f) *Helianthus annum*, g) *Paganelia longifolia*, h) *Gardenia jasminoides*.

Figure 6. Stingless bee worker foraging on a) *Murraya paniculata*, b) *Elaeis guineensis*, c) *Ehretica buxifolia*, d) *Carissa carandas*, e) *Pancratium sp.*, f) *Pureria sp.*, g) *Ixora sp.*, h) *Averrhoa bilimbi*.

Figure 7. Stingless bee worker foraging on a) *Ixora parviflora*, b) *Tagetus sp.*, c) *Artocarpus heterophyllus*, d) *Celosia sp.*, e) *Urena lobata*, f) *Euphorbia heterophylla*, g) *Anthricum andrianum*, h) *Lilium candidum*.

Figure 8. Resin sources on which Stingless bee worker foraging a) *Anacardium occidentale*, b) *Artocarpus heterophyllus*, c) *Garcinia cambogia*, d) *Haevea braziliensis*, e) *Arecaria exelsa*, f) *Mangifera indica*.

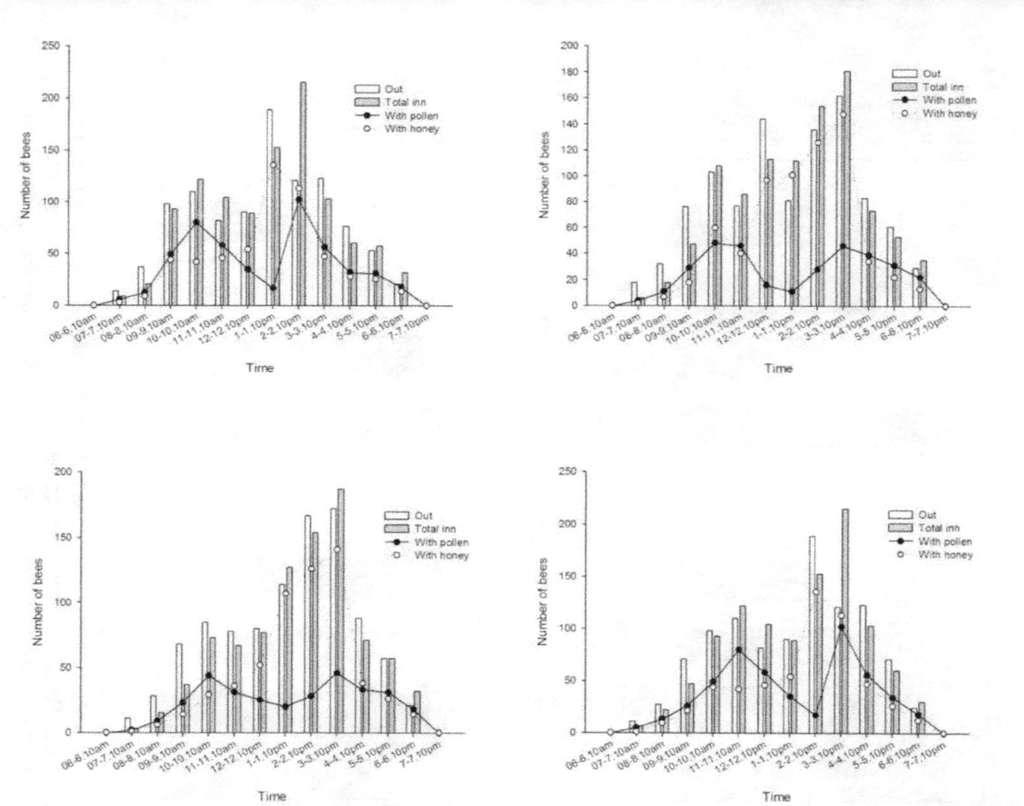

Figure 9. Bee traffic measurements in various timings across day and between four colonies (top left: colony 1; top right: colony 2; bottom left: colony 3; bottom right: colony 4). Measurements include total bees out, total bees in, with pollen and with honey.

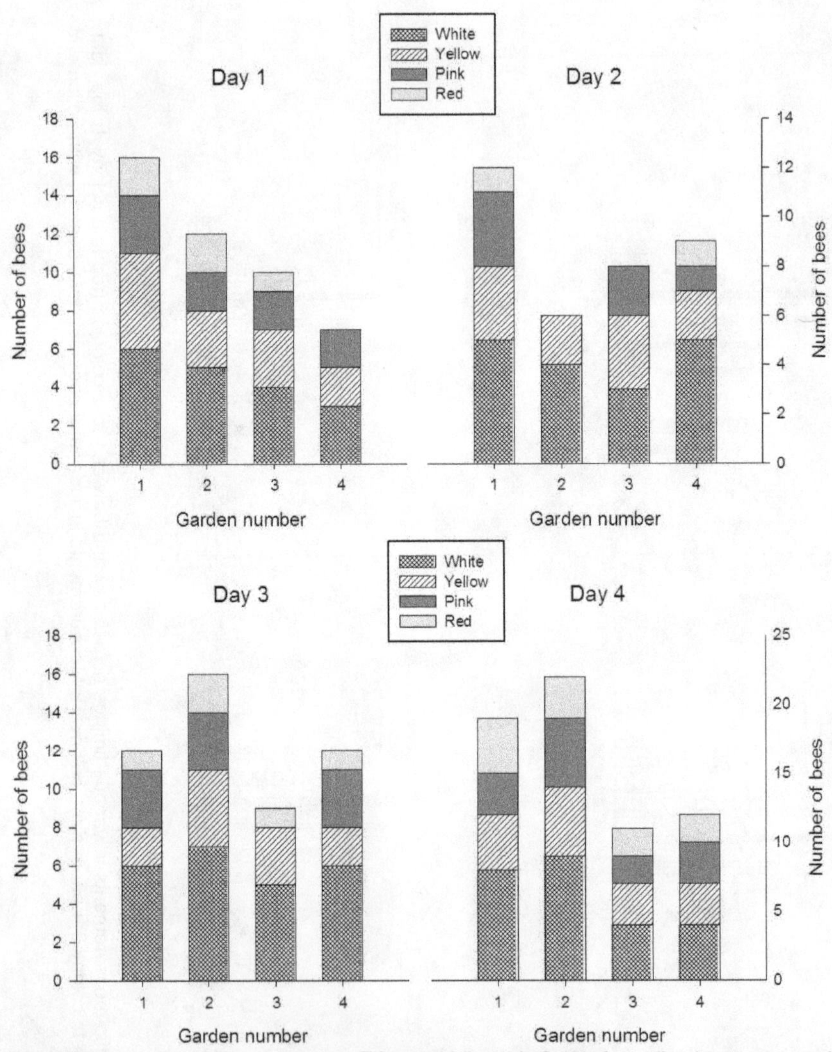

Figure 10. Colour preference among *Trigona iridipennis* Smith for collecting nectar and pollen in rose-moss (*Portulaca grandiflora;* garden 1 and 2) and Ixora (*Ixora coccinea;* garden 3 and 4) in 4 consecutive days (day 1 to 4) in four selected gardens during 2013.

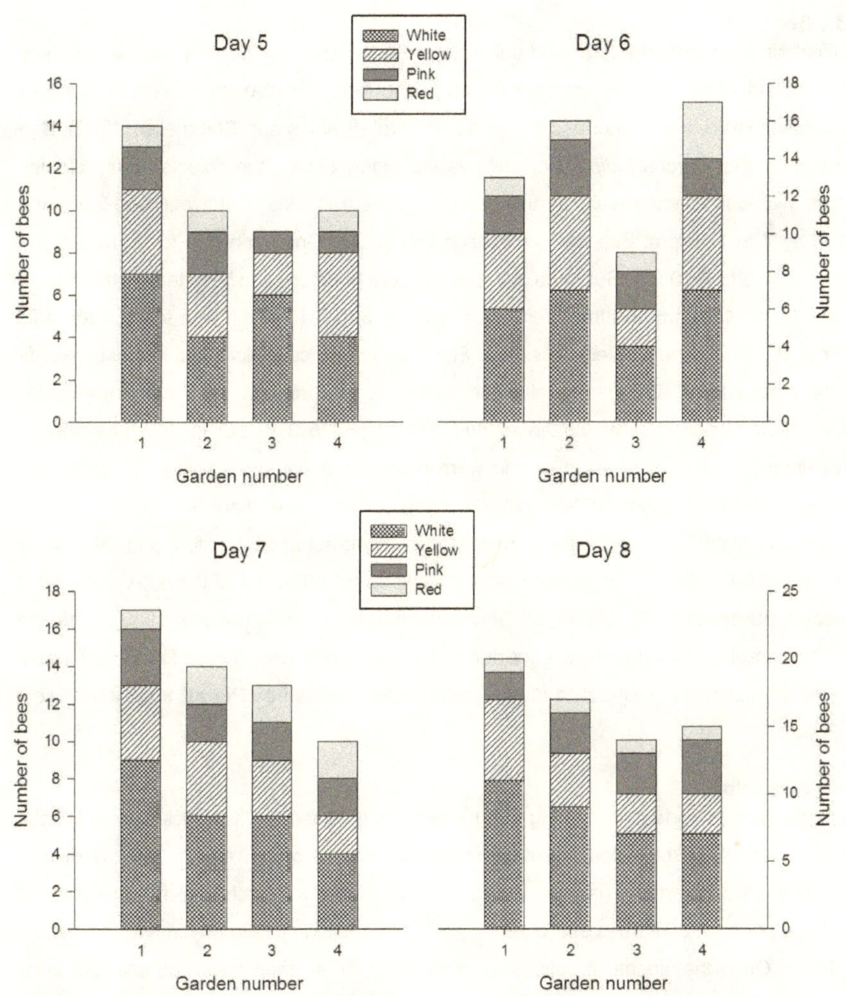

Figure 11. Colour preference among *Trigona iridipennis* Smith for collecting nectar and pollen in rose-moss (*Portulaca grandiflora;* garden 1 and 2) and Ixora (*Ixora coccinea;* garden 3 and 4) in 4 consecutive days (day 5 to 8) in four selected gardens during 2013.

3.8 Bee traffic

Stingless bees collect nectar and pollen to provision their nest. The movement of bees in and out of the nest is not only for collecting food but also for other activities like collecting resin, water and also for waste removal (Kolmes and Sommeijer, 1992). Bee traffic among *Trigona iridipennis* Smith varied widely among the selected four colonies during various time intervals. In all the colonies bee traffic starts between 6 to 7 am and end by 7 to 7.10 pm. Bee traffic exhibited two peaks. One during 10 to 10.10 am and the other 2 to 3.10 pm. But in colony 4 the picture is different, where the morning peak was 11 to 11.10 am and the afternoon peak was 3 to 3.10 pm. It was found that traffic for pollen and honey collection varied. Flight for honey collection was highest towards afternoon where as pollen collection exhibited a forenoon and afternoon peak. Maximum bee traffic was observed during 1.00 pm to 3.10 pm in all four colonies. Maximum number of bee out and in were recorded in colony 1>colony 2> colony 4> colony 3. Pierrot and Schlindwein (2003) reported more than 90% of the pollen collection flights are made in the morning. In favourable season, foraging take place throughout the day but it is most active between 11.00 and 13.00 hrs. A peak flight activity observed at 28°C (Kleinert-Giovannini and Imperatriz-Fonseca, 1986). Variation in the foraging time observed may due to regional variations. It was found that waste removal was more frequent in all the colonies as reported by Pierrot and Schlindwein (2003).

3.9 Swarming

Stingless bees perform swarming due to a variety of reasons which includes size of the nest, strength of the colony, climatic conditions, food storage, and availability of new nesting sites. Swarming usually occurs from November to March, which is the peak of growth period. Scout bees searches for available nesting sites in daytime in the nearby places. Once they locate a suitable nesting site, worker bees clean the site and bring resin, food and other resources to the site. A conspicuous entrance tube was also made by the worker bees in the new site. Unusual activities can be also observed in the parent colony which includes vigorous flight movements in front of the nest which continues for few days. The flight pattern is also quite different which has quick and circular mode of action. Once all the provisions were supplemented in the new site, on a bright day, the worker bees accompanying the Gyne occupy the new nest.

3.9.1 Aggressive swarming

Aggressive swarming was completely different from the above mentioned normal swarming where the flight movements were quite irregular. Two worker bees clutches together and bite each other which finally leads to their fatal death as well as predation from natural enemies. Aggressive swarming happens due to an invasion of a new swamp from a nearby colony which unfortunately selected this nest as their nesting site.

3.10 Drone movement

Drone movements were complexly different from the aforesaid movements where the movements are massive and the head of the bees direct towards the nest entrance. The flight pattern occurs in a forward back ward pattern, and some of the bees found nesting on nearby surfaces like walls, twigs and leaves. Trap net sample collection revealed the bees as drones. The drone movements last up to 4 days in the current study. Only five out of twenty selected colonies showed drone movements.

3.11 Colour preference

Colour preference among *Trigona iridipennis* Smith varied widely showing preference in various colours. White and yellow were the predominant colours for stingless bees. The preference among same colour in a plant variety also differed across various selected days. A clear difference among flower colour selection were observed among the two plant variety (*Portulaca grandiflora* and *Ixora coccinea*).

Table 11. Swarming behavior of *Trigona iridipennis* Smith in selected colonies (20 boxes) during 2013-2014.

Box number	Colony strength*	Number of swarming	Period	Duration (day)	Time
1	S	1	Nov	3	9.00 to 5.00
2	W	0	-	-	-
3	W	0	-	-	-
4	M	0	-	-	-
5	W	0	-	-	-
6	M	0	-	-	-
7	S	1	Jan	4	9.00 to 6.00
8	M	1	Jan	3	9.00 to 6.00
9	M	0	-	-	-
10	S	0	-	-	-
11	S	0	-	-	-
12	S	0	-	-	-
13	M	0	-	-	-
14	S	1	Dec	3	9.00 to 5.30
15	S	0	-	-	-
16	W	0	-	-	-
17	S	1	March	4	9.00 to 6.15
18	S	0	-	-	-
19	S	0	-	-	-
20	S	1	Feb	3	9.00 to 6.00

*0-4 guard bees: weak (W); 5-8: medium (M); above 9: strong (S). Minimum 10 strong colonies

Table 12. Aggressive swarming among *Trigona iridipennis* Smith in selected colonies (20 boxes) during 2013-2014.

Box number	Colony strength*	Number of aggressive swarming	Period	Duration (day)	Time
1	S	0	-	-	-
2	W	0	-	-	-
3	W	0	-	-	-
4	M	0	-	-	-
5	W	0	-	-	-
6	M	0	-	-	-
7	S	0	-	-	-
8	M	0	-	-	-
9	M	1	Feb	3	10.00 to 4.00
10	S	0	-	-	-
11	S	0	-	-	
12	S	0	-	-	-
13	M	0	-	-	-
14	S	0	-	-	-
15	S	0	-	-	-
16	W	1	Jan	2	9.30 to 3.30
17	S	0	-	-	-
18	S	0	-	-	-
19	S	0	-	-	-
20	S	0	-	-	-

*0-4 guard bees: weak (W); 5-8: medium (M); above 9: strong (S). Minimum 10 strong colonies

Table 13. Drone movement among *Trigona iridipennis* Smith in selected colonies (20 boxes) during 2013-2014.

Box number	Colony strength*	Number of drone movement	Period	Duration (day)	Time
1	S	1	Nov	3	10 to 5.00
2	W	0	-		
3	W	0	-		
4	M	0			
5	W	0	-		
6	M	1	March	4	9.45 to 5.00
7	S	0	-		
8	M	0	-		
9	M	1	Feb	3	10.00 to 4.30
10	S	0	-		
11	S	0	-		
12	S	0	-		
13	M	0	-		
14	S	1	Dec	3	9.30 to 5.00
15	S	0	-		
16	W	0	-		
17	S	0	-		
18	S	0	-		
19	S	0	-		
20	S	1	March	4	10.00 to 6.00

*0-4 guard bees: weak (W); 5-8: medium (M); above 9: strong (S). Minimum 10 strong colonies

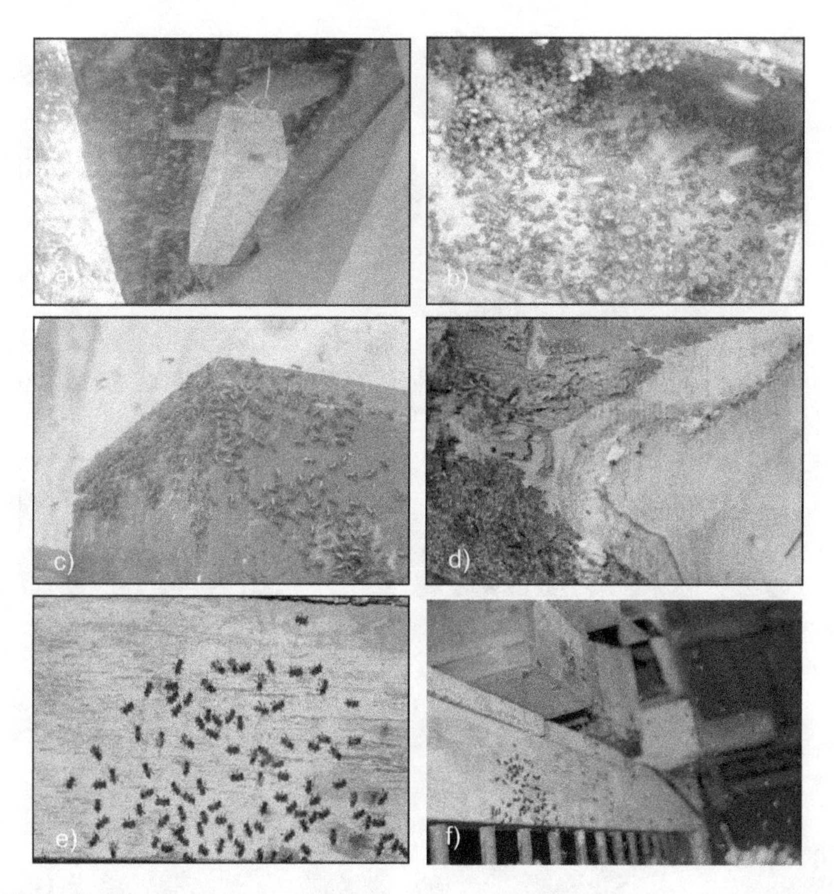

Figure 12. Stingless bees movement a) a swarm entering into a new nest, b) bees killed due to aggressive swarming, c) gyne returns after nuptial flight, d) worker bee collecting resin from *Artocarpus hirsutus* Lam., e) drone bees resting near the nest with gyne, f) drone congregation.

Figure 13. Colour preference among *Trigona iridipennis* Smith for collecting nectar and pollen in *Portulaca sp.* a), b), c) and d). Hives placed in the garden for colour preference study e) and f).

Figure 14. Colour preference among *Trigona iridipennis* Smith for collecting nectar and pollen in *Ixora sp.*

4. Conclusions

Trigona iridipennis Smith shows great diversity in plant selection for dietary as well as resin sources. The shift towards ornamental plants for foraging may be an adaptation evolved in response to human modification of the environment. The bees collect resin from a variety of sources for building nest, its maintenance and also for defence. Bee traffic is found to be related to time, season, and strength of the colony. The bees preferred white and yellow coloured flowers than pink and red. The study also highlights the various food sources of *Trigona iridipennis* Smith in Kerala which can be further explored for flourishing melliponiculture.

Acknowledgements

The first author is grateful for the cooperation of the management of Mar Augsthinose college for necessary support. Technical assistance from Binoy A Mulanthra is also acknowledged. We also thank all the farmers for who cooperate with us during the study.

References

Amano, K., Nemoto, T. & Heard, T. A. (2000). What are stingless bees, and why and how to use them as crop pollinators?-a review. *Japan Agricultural Research Quarterly*, *34*(3), 183-190.

Andualem, B. (2013). Synergistic antimicrobial effect of Tenegn honey (*Trigona iridipennis*) and garlic against standard and clinical pathogenic bacterial isolates. *International Journal of Microbiological Research*, *4*(1), 16-22.

Bänziger, H., Pumikong, S. & Srimuang, K. O. (2011). The remarkable nest entrance of tear drinking *Pariotrigona klossi* and other stingless bees nesting in limestone cavities (Hymenoptera: Apidae). *Journal of the Kansas Entomological Society*, *84*(1), 22-35.

Basavarajappa, S. (2010). Studies on the impact of anthropogenic interference on wild honeybees in Mysore District, Karnataka, India. *African Journal of Agricultural Research*, *5*(4), 298-305.

Biesmeijer J. C., Guirfa M., Koedam D., Potts S. G., Joel D. M. & Dafni A. (2005). Convergent evolution: floral guides, stingless bee nest entrances, and insectivorous pitchers. *Naturwissenschaften*, *92*(1), 444-450.

Breed, M. D., & Page Jr, R. E. (1991). Intra-and interspecific nestmate recognition in Melipona workers (Hymenoptera: Apidae). *Journal of Insect Behavior*, *4*(4), 463-469.

Cameron, E. C., Franck, P. & Oldroyd, B. P. (2004). Genetic structure of nest aggregations and drone congregations of the southeast Asian stingless bee *Trigona collina*. *Molecular Ecology*, *13*(8), 2357-2364.

Camrgo J. M. F. & Pedro S. R. M. (1992). Systematics, phylogeny and biogeography of the Meliponinae (Hymenoptera, Apidae): a mini- review. *Apidologie*, *23*(1), 509-522.

Chinh, T. X., Sommeijer, M. J., Boot, W. J. & Michener, C. D. (2005). Nest and colony characteristics of three stingless bee species in Vietnam with the first description of the nest of *Lisotrigona carpenteri* (Hymenoptera: Apidae: Meliponini). *Journal of the Kansas Entomological Society*, *78*(4), 363-372.

Choudhari, M. K., Haghniaz, R., Rajwade, J. M. & Paknikar, K. M. (2013). Anticancer activity of Indian stingless bee propolis: an in vitro study. *Evidence-Based Complementary and Alternative Medicine*, 2013(2013), 1-10.

Choudhari, M. K., Punekar, S. A., Ranade, R. V. & Paknikar, K. M. (2012). Antimicrobial activity of stingless bee (*Trigona sp.*) propolis used in the folk medicine of Western Maharashtra, India. *Journal of Ethnopharmacology*, *141*(1), 363-367.

Corbet, S. A., Saville, N. M., Fussell, M., Prŷs-Jones, O. E., & Unwin, D. M. (1995). The competition box: a graphical aid to forecasting pollinator performance. *Journal of Applied Ecology*, *1*(1), 707-719.

Cortopassi-Laurino, M., Imperatriz-Fonseca, V. L., Roubik, D. W., Dollin, A., Heard, T., Aguilar, I., Venturiei, G. C., Eardley, C. & Nogueira-Neto, P. (2006). Global meliponiculture: challenges and opportunities. *Apidologie, 37*(2), 275-292.

Danaraddi, C. S. & Viraktamath, S. (2007). Studies on stingless bee, *Trigona iridipennis* Smith with special reference to foraging behaviour and melissopalynology at Dharwad, Karnataka. Master of Science Thesis. College of Agricultural Science. Dharwad.

Danaraddi, C. S., Viraktamath, S., Basavanagoud, K. & Bhat, A. R. S. (2010). Nesting habits and nest structure of stingless bee, *Trigona iridipennis* Smith at Dharwad, Karnataka. *Karnataka Journal of Agricultural Sciences*, *22*(2), 310-313.

David, W. R. (2006). Stingless bee nesting biology. *Apidologie*, *37*(2), 124-143.

Devanesan S., Shailaja K. K., Pramila K. S. (2009). Status paper on Stingless bee *Trigona iridipennis* Smith. All India Co-ordinated Research Project on Honey bees and pollinators, Vellayani Centre,Thiruvananthapuram.pp79.

Drumond, M. P., Zucchi R., Yamane, S. & Sakagami S. F. (1998). Oviposition behaviour of the stingless bees XX. *Plebeia (Plebeia) juliani* which forms very small brood batches (Hymenoptera:Apidae,Meliponinae). *Entomological Science*, *1*(2), 195-205.

Franck, P., Cameron, E., Good, G., Rasplus, J.Y. & Oldroyd, B. P. (2004). Nest architecture and genetic differentiation in a species complex of Australian stingless bees. *Molecular Ecology*, *13*(8), 2317-2331.

Goulson, D., Chapman, J. W., & Hughes, W. O. (2001). Discrimination of unrewarding flowers by bees; direct detection of rewards and use of repellent scent marks. *Journal of Insect Behavior*, *14*(5), 669-678.

Greco, M. K., Hoffmann, D., Dollin, A., Duncan, M., Spooner-Hart, R., & Neumann, P. (2010). The alternative Pharaoh approach: stingless bees mummify beetle parasites alive. *Naturwissenschaften*, *97*(3), 319-323.

Heard, T. A. (1999). The role of stingless bees in crop pollination. *Annual Review of Entomology*, *44*(1), 183-206.

Hubbell, S. P., & Johnson, L. K. (1978). Comparative foraging behavior of six stingless bee species exploiting a standardized resource. *Ecology*, *59*(6), 1123-1136.

Inoue, T., Sakagami, S. F., Salmah, S. & Yamane, S. (1984). The process of colony multiplication in the Sumatran stingless bee *Trigona (Tetragonula) laeviceps*. *Biotropica*, *16*(2), 100-111.

Jaenike, J. & Holt, R. D. (1991). Genetic variation for habitat preference: evidence and explanations. *American Naturalist*, *137*(1), S67-S90.

Jayarathnam K. (1970). *Hevea brasilensis* as a source of honey.J.Palynol.6, 101-103.

Jose, S. K & Thomas, S. (2012). Stingless beekeeping (Meliponiculture) in Kerala. In: Selected beneficial and harmful insects of Indian subcontinent, (Thomas, K.S (2012) ed. LAP LAMBERT Academic Publishing GmbH & Co. KG, Saarbruken, Germany.

Jose, S. K & Thomas, S. (2013). Nest architecture of *Trigona iridipennis*. Proceedings of the National Seminar on Invertebrate Taxonomy. Nirmala Academic and Research Publications, Kerala, India.

Kleinert-Giovannini, A., & Imperatriz-Fonseca, V. L. (1986). Flight activity and responses to climatic conditions of two subspecies of Melipona marginata Lepeletier (Apidae, Meliponinae). *Journal of apicultural research*, *25*(1), 3-8.

Kolmes, S. A. & Sommeijer, M. J. (1992). Ergonomics in stingless bees: changes in intranidal behaviour after partial removal of storage pots and honey in Melipona favosa (Hym. Apidae, Meliponinae). *Insects Sociaux*, *39*(2), 215-232.

Krishnakumar, K. N., Prasada Rao, G. S. L. H. V. & Gopakumar, C. S. (2009). Rainfall trends in twentieth century over Kerala, India. *Atmospheric Environment*, *43*(11), 1940-1944.

Kumar, M. S., Singh, A. J. A. R. & Alagumuthu, G. (2012). Traditional beekeeping of stingless bee (*Trigona sp*) by Kani tribes of Western Ghats, Tamil Nadu, India. *Indian Journal of Traditional Knowledge*, *11*(2), 342-345.

Lehmberg, L., Dworschak, K., & Blüthgen, N. (2008). Defensive behavior and chemical deterrence against ants in the stingless bee genus Trigona (Apidae, Meliponini). *Journal of Apicultural Research*, *47*(1), 17-21.

Lima, S. L. & Dill, L. M. (1990). Behavioural decisions made under the risk of predation: a review and prospectus. *Canadian Journal of Zoology*, 68(4), 619-640.

Marisa, H. & Salni, S. (2012). Red wood (*Pterocarpus indicus* wild) and bread fruit (*artocarpus communis*) bark sap as attractant of stingless bee (*Trigona spp*). *Malaysian Journal of Fundamental and Applied Sciences*, *8*(2), 107-110.

Martin, T. E. (2001). Abiotic vs. biotic influences on habitat selection of coexisting species: climate change impacts?. *Ecology*, *82*(1), 175-188.

Michener, C. D. (2000). The bees of the world. Johns Hopkins University Press, London, UK.

Michener, C.D. (1974). The social behaviour of bees. A comparative study. Belknap, Harvard University Press, Cambridge, UK.

Mohan, R. & Devanesan, S. (1999). Dammer bees, Trigona iridipennis Smith (Apidae: Meliponinae) in Kerala. *Insect Environment*, *5*(1), 79-81.

Nair, M. C. & Nair, P. K. K. (2001). Beekeeping by Kanikkars in southern Western Ghats of Kerala. *Indian Bee Journal*, *63*(1 & 2), 11-16.

Nair, M. C. (2003). Apiculture resource biodiversity and management in Southern Kerala. PhD. Thesis. Mahatma Ghandhi University, Kottayam, 277.

Pavithra, N., Shankar, R. & Jayaprakash. (2012). Nesting pattern preferences of stingless bee, Trigona iridipennis Smith (Hymenoptera: Apidae) in Jnanabharathi campus, Karnataka, India. *International Research Journal of Biological Sciences*, *2*(2), 44-50.

Peters, J. M., Queller, D. C., Imperatriz-Fonsaca, V. L., Roubik, D. W. & Stassmann, J. E. (1999). Mate number, kin selection, and social conflict in stingless bees and honey bees. *Proceedings of the Royal Society of London B: Biological Sciences*, *266*(1417), 379-384.

Pierrot, L. M., & Schlindwein, C. (2003). Variation in daily flight activity and foraging patterns in colonies of uruçu-Melipona scutellaris Latreille (Apidae, Meliponini). *Revista Brasileira de Zoologia, 20*(4), 565-571.

Raju, A. J. S., Rao, K. S. & Rao, N. G. (2009). Association of Indian stingless bee, Trigona iridipennis Smith (Apidae: Meliponinae) with Red-listed *Cycas sphaerica* Roxb. (Cycadaceae). *Current Science, 96*(11), 1435-1436.

Ramanujam, C. G. K., Fatima, K. & Kalpana, T. P. (1993). Nectar and pollen sources for dammer bee (*Trigona iridipennis* Smith) in Hyderabad (India). *Indian Bee Journal, 55*(1/2), 25-28.

Rasmussen, C. (2013). Stingless bees (Hymenoptera: Apidae: Meliponini) of the Indian subcontinent: Diversity, taxonomy and current status of knowledge. *Zootaxa, 3647*(3), 401-428.

Roubik, D. W. (1995). *Pollination of cultivated plants in the tropics* (No. 118). Food & Agriculture Organization, USA.

Roubik, D. W. (1989). Ecology and natural history of tropical bees. Cambridge university Press, New York.514p.

Roubik, D.W. (2006). Stingless bee nesting biology. *Apidologie, 37*(2).124-143.

Singh, R. P. (2013). Domestication of *Trigona iridipennis* Smith in a newly designed hive. *National Academy Science Letters, 36*(4), 367-371.

Smith, B. H., & Roubik, D. W. (1983). Mandibular glands of stingless bees (Hymenoptera: Apidae): Chemical analysis of their contents and biological function in two species of Melipona. *Journal of Chemical Ecology, 9*(11), 1465-1472.

Sommeijer M. J., De Bruijin L. L. M., Meeuwsen J. A. J & Slaa E. J. (2003).Stingless bees nest departures of non-accepted gynes and nuptial flights in *Melipona favosa* (Hymenoptera:Apidae,Meliponini).*Entomologische Berichten, 63*(1), 7-13.

Sommeijer, M. J., Beuvens, F. T. & Verbeek, H. J. (1982). Distribution of labour among workers of *Melipona favosa* F: Construction and provision of brood cells. *Insectes Sociaux, 29*(2), 222-237.

Tereda, Y., Garofalo, C. A. & Sakagami, S. F. (1975). Age-survival curves for workers of two eusocial bees (*Apis mellifera* and *Plebeta droryana*) in a subtropical climate, with notes on worker polyethism in *P. Droryana. Journal of Apicultural Research*, *14*(3), 161-170.

Van Vein, J. W., Sommeijer, M. J. & Meeuwsen, F. (1997). Behaviour of drone in Melipona (Apidae, Meliponini), *Insetes Sociaux*, *47*(1), 70-75.

Virkar, P. S., Shrotriya, S. & Uniyal, V. P. (2014). Building walkways: observation on nest duplication of stingless bee *Trigona iridipennins* Smith. Ambient Science, 1(1), 38-40.

Printed in the USA
CPSIA information can be obtained
at www.ICGtesting.com
LVHW051928170924
791294LV00003B/598